HERBIVORES
in the Food Chain

ALICE B. McGINTY
Photography by DWIGHT KUHN

The Rosen Publishing Group's
PowerKids Press™
New York

To my mother, Linda K. Blumenthal—Alice B. McGinty
To Kylie—Dwight Kuhn

Published in 2002 by The Rosen Publishing Group, Inc.
29 East 21st Street, New York, NY 10010

First Edition

Book Design: Maria Melendez
Project Editor: Emily Raabe
All photographs © Dwight Kuhn

McGinty, Alice B.
Herbivores in the food chain / Alice B. McGinty.
      p. cm. — (The library of food chains and food webs)
Includes bibliographical references (p.   ).
ISBN 0-8239-5753-5 (lib. bdg.)
1. Herbivores—Food—Juvenile literature. 2. Food chains (Ecology)—Juvenile literature. [1. Food chains (Ecology) 2. Herbivores.
3. Ecology.] I. Title. II. Series.
QL756.5 .M38 2002
591.5'4—dc21

                                                                                    2001000152

Manufactured in the United States of America

Contents

What Are Food Chains?

You need energy to live and grow. Your energy comes from the foods that you eat. All living things require food. When one living thing eats another, they form a food chain.

Producers in the food chain make their own food. They use the energy from sunlight to make sugar. Plants are producers. **Consumers** in the food chain cannot make their own food. They must eat other living things. Some consumers, such as rabbits, eat plants. These consumers are called herbivores. Herbivores are known as **primary consumers**. They are the first food-eaters in the food chain. Other consumers, such as

Most animals are part of many food chains. For example, mice eat berries and seeds from plants. Mice are eaten by other animals, such as weasels, cats, and owls. Mice and the animals that eat mice belong to several different food chains. Food chains that are linked together form a food web.

4

wolves, eat other animals. These meat-eating consumers are called carnivores. Carnivores are **secondary consumers** in the food chain. This means that they are the second food-eaters in the food chain. Every food chain ends with **decomposers**. Decomposers, such as mushrooms, are **organisms** that break down the bodies of dead plants and animals.

A meadow vole is an herbivore that is eaten by foxes, snakes, weasels, and other carnivores.

Passing on the Energy

Most herbivores, such as horses, eat different kinds of plants. Some, however, such as the koala, eat only one kind of plant. Koalas eat only leaves from the eucalyptus tree.

The energy from sunlight helps plants make sugar. When a plant grows, some of its sugar is used to make **cellulose**. Cellulose forms the walls of the plant's **cells**.

Herbivores eat plants to get energy from the sugar in them. A horse, for example, uses the sugar in grass to grow and to stay alive. The horse's body must break down the tough cellulose in the plant to release the sugar. The herbivore's body has special organisms inside it that are able to break down the cellulose. Herbivores such as horses have to eat a lot of plants to get enough energy to survive.

6

Horses spend a lot of their time eating. Elephants, the largest herbivores, spend 18 hours every day eating grass and leaves. When another animal, such as a wolf, eats an herbivore, some of the energy from the herbivore is passed on to the wolf and along the food chain.

The tough parts of plants, such as stems, wood, and bark, are made mostly of cellulose. Horses, such as this mother and baby, are able to break down cellulose to get the sugar that is stored in plants.

Tiny Herbivores

All living things are made of cells. Some living things, such as people, are made of many cells. Other living things are made of just one cell. These one-celled organisms are so tiny that you can see them only through a microscope. Many of these tiny creatures live in the sea. In the sea, tiny herbivores called **zooplankton** eat plants called **phytoplankton**. Some of these tiny herbivores are called copepods. Copepods look like little grains of rice with long **antennae**. They use their antennae to collect phytoplankton. Copepods and other tiny herbivores are important links in ocean food chains. They are eaten by many kinds of fish.

Phytoplankton live near the water's surface where sunlight shines through. Zooplankton need to live near the surface, too, to stay close to their food. Zooplankton have special ways of staying near the top of the water without sinking. Copepods, for example, store their extra food as oil drops. These drops of oil help them float near the water's surface.

8

This strange-looking creature swimming among green algae in a pond is a waterflea. Waterfleas are tiny water herbivores. Some waterfleas live in the ocean, and others live in freshwater. Waterfleas filter phytoplankton from the water for food.

Herbivore Insects in the Food Chain

There are over 750,000 kinds of insects in the world! About half of them are herbivores.

Plant-eating insects have special mouthparts to help them eat. Insects such as grasshoppers bite off pieces of leaves. They have biting mouthparts and strong jaws for chewing. Tiny insects called leaf miners tunnel through the insides of leaves. Their mouthparts usually face forward so that they can eat as they dig. Other insects, such as butterflies, suck liquid from plants. Their mouthparts include a tube called a **proboscis**. The insect uses its proboscis like a straw to suck nectar from flowers.

Plant-eating insects are primary consumers in the food chain. They provide food for many secondary consumers.

10

An aphid sticks its proboscis into a plant's leaves or stem to suck out the liquid sap. The aphid pumps extra sap out of its body as a syrup called honeydew, Ants like to eat the honeydew. Aphids are tiny and very common. There can be as many as 5 billion aphids in one acre of forest.

The caterpillar of the monarch butterfly eats bitter-tasting plants called milkweed and dogbane. This makes the monarch taste bitter to any animal that eats it. The monarch's bright colors warn other animals not to eat it.

Grasshoppers use their strong jaws to chew the tough stems and leaves of plants.

The Herbivore's Ecosystem

Scientists called **ecologists** study food chains. Ecologists who study herbivores look at where the herbivore lives, what it eats, and everything that surrounds it. This is the animal's **ecosystem**.

Herbivores compete for food in every ecosystem. To survive, each kind of herbivore must find a place in the ecosystem. In the grasslands of Africa, giraffes eat leaves from the treetops. The giraffe's long neck reaches leaves that are too high for other herbivores to eat. In the desert ecosystem, camels use their tough, rubbery lips to eat prickly plants that other herbivores are unable to eat.

When this wood frog tadpole gets older, it will grow legs and live on land. As a frog, it will eat insects instead of water plants.

Slugs eat the leaves, stems, fruits, and flowers of plants. They scrape off pieces of the plant with tongues that are covered with thousands of tiny teeth.

This shows a close-up of a tadpole's mouth. Tadpoles scrape algae from the bottom of the pond or stream where they live.

Herbivore Birds

Birds must eat more food than any other animal their size. They also must digest their food quickly to get the energy they need. Some birds can digest their food in half an hour.

Birds are important herbivores in a food chain. Herbivore birds eat foods such as seeds, nuts, fruits, and nectar. Looking at a bird's beak gives us clues about what kinds of foods it eats. Birds that eat nuts and seeds, for example, have short, strong beaks. They use their beaks like nutcrackers to crack open the hard shells and seeds. Birds do not have teeth. Inside their bodies, they have organs called gizzards that grind food. Birds may swallow small stones to help their gizzards grind food. Birds also have a food storage area called a crop. This allows the bird to eat a lot of food and then fly to a safe place to digest it.

Hummingbirds have very long, thin beaks. They use their beaks to reach into the center of flowers and drink nectar.

This cedar waxwing eats nuts, seeds, and fruits to get the energy that it needs to fly.

This yellow-headed amazon parrot's beak is strong, short, and hooked. It is the perfect tool to crush the hard seeds the parrot likes to eat.

Mammal Herbivores

Mammals are animals that feed their babies with the mother's milk. Many, although not all, mammals are herbivores. Some mammal herbivores, such as cows, have stomachs with four sections! When a cow swallows grass, the grass goes to the first section of the cow's stomach. There, it mixes with tiny organisms, including **bacteria**. The tiny organisms break down the plant cellulose into a paste. Later the cow brings wads of this paste back into its mouth. These wads are called cud. The cow chews the cud to break down the cellulose even more. When the cow swallows the cud, it is digested in the three other sections of the cow's stomach. It takes three to five

Herbivores such as deer swallow their food with very little chewing. This allows the animal to eat a lot of food quickly. Early in the morning, the deer feeds on grass, bark, leaves, and twigs. During the day, the deer finds a safe place to hide from its enemies and to digest its food.

16

days for most cows to digest their food. Other herbivore mammals, such as horses, zebras, and elephants, have only one stomach. They too have tiny organisms in their bodies to help them digest cellulose.

The first section of the cow's stomach is the largest part of the stomach. The other three sections of a cow's stomach digest the food after the animal swallows its cud. A cow's stomach can hold about 35 gallons (133 liters) of digested food. Other animals that chew their cud are sheep, goats, giraffes, and camels.

An Herbivore's Teeth

A cow has eight sharp incisors on the bottom of its mouth, but like the deer, it does not have incisors on the top of its mouth. Instead, the cow has a tough pad on the top of its mouth. A cow also has 24 molars for chewing and grinding tough food.

Herbivores have teeth that are shaped to help them eat their food. Beavers have long front teeth for chewing wood and bark. These front teeth are called **incisors**. The beaver's incisors grow all the time. The beaver must gnaw on things to keep its incisors from getting too long. Grass-eating, or grazing, herbivores have incisors, too. Some grazers, such as deer, only have incisors on the bottoms of their mouths. They have a tough pad of skin where their top teeth would be. The bottom incisors work like an axe to chop off plant stems against the pad of skin. Elephants eat soft leaves instead of grass. Their teeth are not as hard as those of the grass eaters.

18

Deer have molars in the backs of their mouths. Molars are flat, wide teeth with ridges. They are used to crush and grind food.

Deer have 32 teeth, including 6 incisors and 24 molars. A deer grinds its cud by moving its jaws from side to side and crushing its cud with its molars.

Producers and Herbivores

In order to break down the cellulose from the plants that they eat, rabbits must eat their food twice. The rabbit eats its droppings so its food can be digested a second time. Although this may seem strange, it allows the rabbit to survive by getting energy from plant foods.

In every ecosystem, producers and herbivores must be balanced. If there were too many herbivores, they would eat all the plants. To keep a balance, carnivores must eat the herbivores. When deer and opossums were brought to the country of New Zealand, for example, there were no predators to eat them. The deer and opossums ate all of the plants in the area where they lived. They turned the land into a desert. On the other hand, if there were not enough herbivores, then producers would become too popular. This happened when people brought prickly pear cacti to Australia. There were no herbivores in Australia to eat the cacti.

The cacti became too popular and grew all over the Australian grazing fields. The Australians had to bring in moth larvae to eat the prickly pear cacti.

This cottontail rabbit serves as food for many animals, such as hawks, wolves, and foxes. Rabbits pass the energy they get from plants to other animals through food chains.

Protecting the Ecosystem

A bee brings pollen from one flower to another as it drinks nectar. This allows each flower's plant to make more plants. A squirrel buries seeds and acorns to store them as food for the winter. Sometimes the squirrel forgets to eat all of its buried food. Many of the forgotten seeds grow into new trees. Though these animals may not realize it, they are protecting their ecosystem. By helping new plants grow, the herbivores ensure that they will have food for the future. They also are making sure that their children will have enough food. People also need to protect their ecosystems. If we work hard to take care of our world, we too will provide a good future for our children.

Glossary

antennae (an-TEH-nee) Two feelers attached to an animal's head that the animal uses to touch, taste, hear, and smell what is around it.

bacteria (bak-TEER-ee-uh) Tiny living things that can be seen only with a microscope.

cells (SELZ) The many tiny units that make up living things.

cellulose (SEL-yuh-lohs) The substance that makes up the main part of the cell walls of plants.

consumers (kon-SOO-merz) Members of the food chain that eat other organisms.

decomposers (dee-kum-POH-zerz) Organisms, such as fungi, that break down the bodies of dead plants and animals.

ecologists (ee-KAH-luh-jists) Scientists who study the way living things are linked with each other and with Earth.

ecosystem (EE-koh-sis-tum) The way that plants and animals live in nature and form basic units of the environment.

incisors (in-SY-zurz) An animal's front teeth that are used for cutting.

organisms (OR-geh-nih-zehmz) Living beings made of dependent parts.

phytoplankton (fy-toh-PLANK-tun) An ocean or freshwater plant made up of one cell.

primary consumers (PRY-mehr-ee kon-SOO-merz) Members of the food chain that eat plants, which makes them the first link of consumers in the food chain.

proboscis (pruh-BAH-sis) A tubelike mouthpart that insects use to suck in liquid food.

producers (pruh-DOO-serz) Plants or tiny organisms that use energy from sunlight to make their own food.

secondary consumers (SEH-kun-der-ee kon-SOO-merz) Members of the food chain that eat plant-eating animals, which makes them the second link of consumers in the food chain.

zooplankton (zoh-uh-PLANK-tun) Tiny animals that float freely in water.

Index

Web Sites

Due to the changing nature of Internet links, PowerKids Press has developed an online list of Web sites related to the subject of this book. This site is updated regularly. Please use this link to access the list:
www.powerkidslinks.com/lfcfw/herb/